THE HORSE

A Firefly Book

Published by Firefly Books Ltd. 2017

Copyright © 2017 Firefly Books Ltd.
Original French-language © 2016 Editions Glénat

All rights reserved. No part of this publication may be reproduced, stored in a retrieval system, or transmitted in any form or by any means, electronic, mechanical, photocopying, recording or otherwise, without the prior written permission of the Publisher.

First printing

Publisher Cataloging-in-Publication Data (U.S.)

Library of Congress Cataloging-in-Publication Data is available

Library and Archives Canada Cataloguing in Publication

Stuewer, Sabine
[Cheval, sa nature révélée. English]
　　　The horse : its nature, revealed / photographs, Sabine Stuewer ; text, Emmanuelle Brengard.
Translation of: Le cheval, sa nature révélée.
ISBN 978-1-77085-959-3 (hardcover)
　　　1. Horses--Pictorial works. 2. Horses--Behavior. I. Brengard, Emmanuelle, writer of added text II. Title. III. Title: Cheval, sa nature révélée. English.

Published in the United States by
Firefly Books (U.S.) Inc.
P.O. Box 1338, Ellicott Station
Buffalo, New York 14205

Published in Canada by
Firefly Books Ltd.
50 Staples Avenue, Unit 1
Richmond Hill, Ontario L4B 0A7

Translator: Francisation InterGlobe

Printed in China

 We acknowledge the financial support of the Government of Canada

Originally published as
Le Cheval Sa Nature Révélée
The Cheval Magazine brand is a property of PrestEdit
© Cheval Magazine
© Editions Glénat for the book adaptation, 2016

THE HORSE

Its Nature, Revealed

PHOTOGRAPHS BY **SABINE STUEWER**
TEXT BY **EMMANUELLE BRENGARD**

FIREFLY BOOKS

INTRODUCTION	6
AMONG THE HAREM	**8**
THE STALLION	15
MARES AND FOALS	45
THE NUANCES OF FRIENDSHIP	**72**
HIERARCHY	79
PARTNERSHIPS	107
GALLOPING TO FREEDOM	**140**
FINELY HONED SENSES	147
AT THE HORSES' PACE	175

This Paso Fino, a South American breed, is sheltering from the rain under trees. Horses, however, aren't made of sugar and can easily tolerate a little rain. Their thick coats and mane keep them comfortable in colder, wetter weather. After they shed their coat in spring, however, they will seek shelter from heavy rain and cool temperatures, particularly when it's windy. In fact, horses are irritated by cold drafts more than by rain, when they will shelter from windy conditions in more heavily wooded areas or behind hedges.

Next pages: A group of Icelandic horses pound their hoofs into the ground in unison. By observing horses' temperaments and behaviors, we can gain insights in the equine character. Horses will however (and thankfully) remain at least in part creatures of mystery.

INTRODUCTION

As it leaves its stall and enters a field, a foal transforms into a slightly different creature. Suddenly, a gust of wind sweeps a few leaves from a nearby tree into the magnificent steed's field of vision. It immediately reacts, stepping back, pivoting on its hind legs and frantically galloping away. After a few strides, it stops, turns around to make sure nothing is chasing it and then snorts loudly before starting to graze. We humans, whether we live close to horses or prefer to love them from afar, are usually quite surprised to observe such unpredictable behavior in a horse. In order to better understand and anticipate horses' behavior we need to learn how to decode it.

Horses are large and powerful animals, but they can also be gentle, particularly with humans. They are extremely sensitive creatures, and act mainly on instinct and by reflex. While training can channel their behavior, it can never truly restrain them. Understanding equine behavior is the surest way of becoming a genuine horse lover. Through Sabine Stuewer's beautiful photographs, this book reveals horses' secrets. These photographs show that, more often than not, horses are simply reacting to stimuli that trigger certain reflexes, such as fleeing.

Most importantly, we should avoid falling into anthropomorphism. A horse never acts abruptly, steps away or gallops off simply to annoy us. That's a human concept. Similarly, what we call "love" and "friendship" between horses should not be interpreted from a human point of view but rather an equine one: there are certainly affinities between particular horses, but these relationships are generally the result of partnerships borne out of necessity, such as for survival, hygiene and the like.

In order to assimilate these varied ideas about equine behavior, we must have an open mind, a keen sense of observation, a thirst for knowledge and, most importantly, a great deal of patience. Horses only reveal their true nature to those who take the time to understand and love them as equines.

The greatest equestrian masters have maintained that such an understanding of horses is required in order to facilitate the true evolution of equestrianism, which is in line with loving and respecting the horse. In the 20th century, the famous Portuguese trainer Nuno Oliveira (1925–1989) relentlessly promoted this view of horses, and one of his many lessons can be put into practice by all: "I ask all equestrians who train their horses to take a good

A harem of horses has nothing to do with the chambers where sultans of the Ottaman Empire kept their wives and concubines. The term is used to designate a relatively small family of horses. The thousands-strong galloping herds of the Wild West plains filmed by Hollywood directors were rallied by humans — cowboys in this case — and were made up of hundreds of harems. In the wild, a breeding male will generally be accompanied by one to three mares, the foals born that year and the young that are two to three years old. The average family is thus only made up of about 10 horses. This relatively small social structure is easier to manage: the stallion expends less energy protecting and keeping just a few females. The harem provides a certain form of stability, but it also evolves, especially once the young horses leave the "nest."

Young males generally leave the harem around two years of age. They either leave of their own accord or are driven out by the breeding male. They will then form temporary groups of non-breeding males, made of up about 15 stallions. Young females are generally "seduced" by a young stallion seeking his own harem. But there is no romance in this story and no Prince Charming, even if the stallion is a magnificent white steed! For horses, this is simply the expression of their natural need to reproduce: mares that leave have reached sexual maturity (around three years old) and willingly follow a stallion that has come for them. Mares sometimes demonstrate a strong attachment to their harem and refuse to follow another stallion. Luckily, such cases are rare. These random meetings are essential to ensuring genetic diversity. Moreover, scientists have observed that stallions drive their sons out of the family, and young mares will only stay with their mothers if the head of the family has changed. This happens when an elderly stallion loses his harem to a younger male.

The herds of non-breeding males, often referred to as bachelor herds, are also constantly evolving. A young male will leave the bachelor herd if he is able to form his own harem, while an older stallion may join after having his harem taken by a younger male. Some males may even remain as "bachelors" their entire life; not all of them are able to get a mare's attention, and some are just not strong enough to fight another stallion and take his place. Some fillies (mares that have not yet reached sexual maturity) have also been observed within bachelor herds.

Against all odds, and despite the instinct that leads horses to live in herds, some horses remain solitary and live alone for a certain period of time. It may be an old stallion that has not yet found his place within a bachelor herd, a young stallion searching for a harem or an old mare who can no longer breed.

Opposite: A tender moment between a Norwegian Fjord mare and her foal. One can feel the gentleness of a mother with her little one; the attention she gives brings comfort and safety. With its still fluffy mane and its adorable face, this foal really seems to enjoy its mother's affection.

AMONG THE HAREM

When the sun is at its peak, these purebred Lusitano mares (a Portuguese breed) usher their foals into the leafy shade to shield them from the intense heat. Under the canopy of trees, some take a moment to rest while others watch on high alert, turning their eyes and ears toward intriguing objects and noises. If danger is imminent and a predator is detected, a horse will sound the alarm and the herd will flee to a more secure area.

THE STALLION

Every harem is led by a stallion who watches for predators and other stallions — his main rivals for his harem. If he sees or hears anything suspicious, he sounds the alarm. His posture changes, his muscles contract and he raises his head and points his ears forward to hear better. If danger is near, he will loudly blow air out his nostrils, and on his signal, the entire group will gallop as quickly as possible away from the threat. Every member of the group may behave this way, but since the stallion is the most vigilant observer, he most often gives the signal to flee. However, he always fights alone against a rival stallion who is trying to appropriate one of his females. These rival stallions generally come from a hierarchical group of non-breeders. Within these groups, males simulate fights, training themselves for the real thing conflicts that await them. But these duels are not systematic — sometimes only a few spirited postures (such as high trotting or mane raising) are enough to intimidate the opponent and cease hostilities. However, when real fights do happen, they can be violent, and most males bear the marks of these blows and bites on their coats. In the wild, a single injury can have serious consequences. If a stallion has injured one of his limbs, he will not be able to travel, and if he has injured his jaw, he will not be able to eat. In both cases, the injuries will likely be fatal.

Relationships between mares are, mostly out of necessity, friendlier. The stallion will usually have one or two favorite females, and he will seek them out when resting or for mutual grooming. To groom each other, two horses scratch each other in the areas of their body that they cannot reach themselves. Stallions usually keep their distance while mares are giving birth, as some females can be unsociable with him just after giving birth. Stallions are not usually aggressive toward their young, but a mare may refuse his advances for a time, before once again mating with him and bearing young.

Opposite: The chestnut coat of this purebred Lusitano stallion shines like silk in the sunlight. This impressively muscled male is a magnificent example of his breed. His forelock and mane bounce to the rhythm of his galloping gait. He is calm though, and we can glean from his eyes that he is not running away in fear.

This purebred Spanish bay stallion, hailing from the Iberian Peninsula, rears up, showing his strength. By pushing up from their hind legs, horses can raise themselves almost vertically into the air and hold that position for a few moments. They usually do this to either intimidate a rival or when simulating a battle with a member of their herd. In this photograph, even though the stallion's ears are pointing back, he does not seem to be challenging a rival: his head muscles are not contracted, and his eyes look rather calm.

This rearing Arab-Barb stallion, however, does appear to be preparing for a fight. Rather than the gesture itself (throwing its front leg forward), it is the flexing of his muscles that signals his intentions to the patient observer. His ears are flat against the nape of his neck, his nostrils are flared, his mouth is tense and slightly agape and his gaze has hardened. A rival looking to steal a mare away from his harem could easily be dissuaded by such an intimidating posture.

Opposite: On average, stallions spend twice as much time monitoring their surroundings as other members of the harem. The horse's vigilance is constant during this watch. By observing these leopard-spotted Curly (top) and (at bottom) piebald Mangalarga Marchador stallions, we notice that their postures are rather relaxed. Their ears face forward and their necks are raised, maximizing their hearing and vision to ensure that their surroundings are safe. Logically, the stallion faces into the wind, so that he can smell potential predators approaching.

Right: This Mangalarga Marchador stallion (a South American breed) is agitated and his muscular tension is clearly visible. His ears move in all directions, his mouth is ajar and the muscles around his eyes are tightened. He watches behind him, the whites of his eyes plainly apparent. The movement of his hairs bely his body's motion with his neck rounded out. Did he just huff? This sound acts as a warning signal, informing other members of the group of imminent danger.

This Friesian stallion is particularly elegant with his rounded physique, undulated coat, full fetlocks and high trot. This so-called "Baroque" horse is a Dutch breed characterized by its ebony coat, where not a single white hair shows through. Stallions choose to trot during play, when intimidating an adversary or when parading for a mare. He may also trot when fleeing a perceived threat, but this particular stallion's tranquil demeanor does not seem to indicate stress or fear.

A splendid mane drapes over the horse's neck, floating in the slightest movements of a breeze. These Black Forest (above) and Haflinger (right) horses, hailing from Germany and Austria, respectively, have washed-out and near-white crests, giving them a certain touch of beauty and offering an incredible contrast to their burnished and golden chestnut coats. However, rather than elegance, horses care more about the efficacy of their coats' protection against insects.

Next pages: A stallion energetically trotting on the moors of its homeland. The Icelandic horse is the only equine breed of its island, and it's particularly resistant to the harsh elements of the region. As a special advantage, this horse has not three, but five gaits. On top of pacing, trotting and galloping, it can also amble (lateral bipedal movement, in contrast to trotting, where the horse moves diagonally) and perform the tölt (a four-beat amble). Both gaits are innate to the breed.

> " Reigning in one's own family is no less difficult than governing an entire province. "
>
> *Tacitus, Roman senator and historian (58-120 AD)*

Stallions develop a more imposing musculature than do mares. In particular, a stallion's neck is thicker and more rounded, presenting a proud appearance. This cream-coated Barb horse is breathtakingly beautiful: his immaculate coat, belying an impressive set of muscles, shines like a pearl in the sunlight. He presents an attitude of attentiveness to his environment, yet remains calm.

The most striking feature of a stallion is his propensity to "show off." Sabine Stuewer was fortunate enough to take this picture, which not only shows the extraordinary beauty of this gray Spanish purebred, but also showcases his incredible charisma. Indeed, his undulated mane gives him a majestic appearance, but it is his attitude in particular which demonstrates that he is the boss of the herd!

Opposite: This brown bay Mangalarga Marchador has adopted a posture typical of the stallion: he trots around, his neck as if pulled toward the ground, his head swinging up, down and to the sides, his ears flat on the nape of his neck and his eyes staring straight at his target. Stallions adopt this "herding" behavior when they seek to unite their harems and get them to move forward. This "threatening" practice can be viewed as an order, and mares tend to follow it without second thought.

Above: This chestnut Welsh stallion has adopted yet another form of intimidation: he energetically trots with his legs up high, his neckline raised, his tail up and his upper lip stretched downward. This stallion is parading and, according to the situation, he could either be trying to impress an opponent or to seduce a mare. In both cases, the goal is to show off his assets.

Two Friesian stallions rear before each other. The two males are dedicated to the showdown. In contrast with other animal species, horses do not fight to maintain territory. Rather, they challenge each other's hierarchical dominance. The struggle between two males aims to sort out their positions. A typical scenario of this type begins with the arrival of a young stallion who has come to "steal" a mare from an older male in order to form his own harem. The reigning stallion must defend his mares, and the fight only ends when one of the two males shows signs of submission. The loser is then chased away, the victor baring his teeth menacingly.

The struggle between males

Fights between stallions are impressive and often last a long time — hours, even days. Of course, the fighters do take breaks to recuperate. Males put every ounce of their energy into these bouts, which can sometimes lead to serious injuries. And if the trauma is too great, the horse may die — injuries can become infected, a broken limb can limit or entirely prevent the possibility of movement and a broken jaw will prevent the horse from feeding itself.

In a fight between a young stallion and an older one, the advantage often goes to the horse with the most experience: it is not his first battle, and he has certainly had the time to hone his technique. However, a stallion will lose strength, agility and energy with age, and a young challenger may be able to dominate and oust him as a result. In this case, the younger male may take over the loser's entire harem.

During fights, combatants make angry squeaking and "shouting" noises. The males try to bite each other with their mouths wide open and their teeth bared. And if the teeth don't tear flesh, they may still inflict deep wounds nonetheless. The coats of wild stallions are replete with numerous scars that show their bravery in battle.

As if taken by a menacing dance, the horses turn around on themselves to kick each other. They rear and throw their entire weight against an opponent, who may suffer greatly if he can't dodge the kick. Throwing a kick from the fore or the rear is done with force and precision. The hits are brutal and the aggression shown is not a game: males are looking to reproduce and fights can be quite common, especially in the spring when mares are in heat. Every fight is different. Its course changes according to the protagonists' ages, the season and so on. As such, there's no typical belligerent scenario and a fight can stop very quickly, even if it is extremely violent — as if the two boxers had suddenly decided to opt for a draw.

Opposite: Not every meeting between males ends in combat. Sometimes, the sole use of intimidation postures is enough to impress the adversary. A confrontation between two stallions often begins with the same kind of posture presented by this bay Paso Iberoamericano: pawing at the ground with his foreleg and rounding out his neckline to show his strength.

Above: This chestnut-coated Peruvian Paso is also pawing at the ground with his foreleg, raising his head up high to prove his might. At the same time, he swings his neck and swishes his tail, moving his hairs in order to intimidate his opponent. If these postures are not enough, both stallions will begin to trot. They will approach one another to smell and to judge whether a fight will happen or not.

In the foreground, this chestnut-colored Welsh stallion, recognizable by his powerful musculature, arched neckline and parading attitude, is looking to seduce a mare from his harem. She has the final decision on the matter. When she is in heat, she will show interest in the stallion. He will then try to woo her, but if she isn't ready, she may refuse his presence and even threaten him with a kick or a bite, especially if he's insisting at the wrong time.

Next pages: This chestnut mare with a large white mark on her head was not receptive to the stallion's advances. As a result, he will see if another one of the mares in his harem is more inclined. When a nearby mare is in heat, the stallion will produce a flehmen response: he curls back his lips in order to sense the female's pheromones and even her estrogen levels, which indicate her current cycle.

Reproduction

By sniffing the female's genitals, the male checks whether the mare is actually inviting him to copulate. The mare's calm and steady demeanor seems to prove him right. She's a brood mare — that is, she is still flanked by her foal. The latter is perhaps the offspring of the preceding year's mating. Whether as its son or daughter, this foal immediately submits to the stallion by exhibiting a behavior known as snapping: the foal strains its neckline and snaps its jaw as if chewing with its mouth wide open. At the same time, it flattens its tail against its rump and lightly bends its front legs to show its acceptance of the male's dominance.

The stallion is excited and continues to court the mare, leaving no doubts as to the intentions of each horse. When the mare is ready for intercourse, the stallion heads for the female's dock and mounts her by standing on his hind legs, placing his forelegs haphazardly around her body. Sometimes, several attempts are necessary before mating can fully occur.

THE STALLION

This little Mangalarga Marchador foal is resting on a cool spring morning, lying down with its legs folded under his body. Is that the mother grazing beside it? The little one shows its submissiveness, its mouth appearing to chew on air (a behavior known as "snapping"). This is an unusual posture for a foal with its mother. Generally, a young horse will reserve this attitude for a dominant member of the group, to which it wants convey the message, "I'm small and fragile. Please don't hurt me."

The foal's first grooming begins with its mother. These mutual grooming sessions reinforce the bond between horses, as well as serving a hygienic purpose. When molting, horses get rid of dead hairs by scratching each other with their teeth.

Above: Although mares recognize their foals immediately after giving birth, it takes the foal a few days to recognize its mother. But nature helps it along: other mares will refuse to let it feed, and its mother calls it back when it strays a little too far. Once this reciprocal bond is formed, it's very strong, and one of their favorite moments is suckling time. During the first six months of its life, the foal, just like this chestnut Mangalarga Marchador, will nurse more than 40 times a day and night in order to build its strength. As it grows and begins to feed on grass, the suckling is reduced to about 20 times a day.

Opposite: Mom really is the best bodyguard there is — even against insects! With its thin mane and small tail, this young bay Mangalarga Marchador is unable to repel the stinging insects that harass horses all spring and summer. One of the best solutions is to take cover behind its mother's rump so that her long and bushy tail can chase the insects away for the foal.

When the foal needs some rest, this Mangalarga Marchador mare stays close and grazes nearby. While adult horses rest according to a steady sleep cycle consisting of light and deep phases, foals require longer phases of deep sleep in order to grow. When sleeping lightly, horses generally adopt the "cow's" position, with their legs folded underneath their body and only their nose touching the ground.

This splendid Paso Fino mare has given birth to an ebony-coated foal. The standards of this South American breed accepts all colors. Without getting into the details of equine genetics, we can assume that the breeding stallion and/or its mother's ancestors had dark coats. Yet, there are certain light-colored horse breeds that continue to produce foals with black coats. Such is the case of the Slovenian Lipizzaners and of the Camargue horse hailing from the Rhône delta in France. These horses are born with black skin and hair, which gets lighter as they age.

Next pages: In this group of Connemara horses — ponies native to the eponymous region of Ireland — a gray mare flanked by her foal looks older than the other. Her more rounded shape may indicate that she is pregnant. A mare can provide milk for her foal even if she is pregnant with another. However, once the latter is born, the older foal will have to yield to its little brother or sister. Its mother will no longer allow it to suckle, and her attention will be captivated by the newborn. In this way, weaning is an entirely natural process.

> **If freedom is dazzling, the family is reassuring.**
>
> Robert Choquette, *Moi, Petrouchka: Memoirs of a Twenty-Two-Year-Old Cat,* 1980.

MARES AND FOALS

Educating the foals

We can tell not only that these Paso Fino foals stay close to their mothers, but that they also benefit from the tutelage of other mares, since they often move together with their offspring in tow. The agile foals have no problem following the group's movements through all kinds of terrain, and they valiantly accompany the adults on all their travels.

The mothers are responsible for educating the foals. Here, during a good gallop with another mare and foal, a Connemara mare shows an unmistakable attitude with her foal: she turns her head toward it, flattens her ears slightly, and shows it that she is unhappy with its behavior. Is she telling it to stay at her side for more protection, just like the smaller foal with its mother, and to not stray ahead?

Tenderness or a bit of tough love? By pushing her head against her foal's body, this Icelandic mare prevents it from venturing where she does not want it to. Yet, her attitude is rather tranquil and the foal responds just as placidly: its ears face forward and its eyes are peaceful. The foal first assimilates its fellows' body language from its mother, and then from other members of the group. Bit by bit, it will learn how to forge its place within the harem hierarchy.

The young horse's schedule is pretty simple: suckle, sleep and play. And the first partner in play is always its mother. Although this Mangalarga Marchador mare would like a little rest, her foal only wants one thing: to have fun! It scratches its mother's back with its foreleg. Does it think it can convince her to play? Is scratching at her back enough to amuse the foal? In any case, Mom has decided that the games end now! As any good mother, she teaches her child to respect its elders' — and others' rest.

Opposite: With her ears laid back, she shows a threatening attitude to put the little rascal in its place. It immediately understands that it's gone too far, puts on a sheepish look and stops bothering her.

Curiosity is a lovely shortcoming, since it facilitates the foal's education. Every one of these foals is curious — whether the Spanish purebred (mouse gray coat), the German Sport Pony (bay coat), the Curly (spotted chestnut coat with characteristic frizzy hairs) or the Mangalarga Marchador (bay coat) — and they spend lots of time exploring their surroundings. By imitating the adults, especially their mothers, foals learn to find the grasses that won't make them sick. Of course, they are interested by more than just grasses. A piece of wood, a frog near a mare — anything can pique their interest. They sniff at it, touch it with the tip of their nose and, eventually, nibble at the object, sometimes shaking it around or pawing at the ground near it.

Above: Play, as seen with these Mangalarga Marchadors, is a very important part of a foal's life. Foals will jostle, nibble and climb on their "opponent," among other things. Play time is a great outlet that allows foals to forge social bonds and to find their place among the group's hierarchy. At the same time, play is also important for the physical development of young horses, since it allows them to build muscle — all while having fun!

Opposite: These dark-coated foals are Morgans, a breed originally from the United States, whose name comes from the founding stallion, Justin Morgan. The young ones spend almost two-thirds of their time having fun with foals their own age. Following the observations of ethologists, it was noticed that females generally preferred chasing games whereas males tended to playfight more often — games that simulate their future life as stallions.

Foal games

A young Friesian, recognizable by its black coat, nears a bay foal who does not seem to appreciate the approach very much. However, once they give a sniff to recognize each other and understand one another's intentions, the games can begin. The bay foal rears without hesitation — ears forward, head muscles relaxed and forelegs folded, showing excitement at the idea of play.

For these very active foals with unbridled energy, rearing is one of their favorite games, and they will even climb on top of each other, putting one or both of their forelegs on their playmate's back. Foals generally have the same strength as each other, and even if their hits are sometimes quite hard, the risk of injury at their age is minimal.

MARES AND FOALS

Two young Tinkers on a frantic chase. In the foreground, the foal seems to be lagging a little behind its opponent, until it's finally able to give a good shove of its hind legs to propel itself forward and catch back up to its playmate. In the end, it easily passes its companion, and using its momentum, provocatively throws its posterior at the loser. An incitement to continuing the chase, maybe?

2
THE
NUANCES
of friendship

In order to live peacefully as a group, information must be exchanged. Horses use all the tools at their disposal to be understood by their companions, as well as by humans (if the latter would open their eyes a bit). To obtain information about the nature of horses, we must look at their entire bodies, the way they move and so on. Indeed, horses are not very "talkative" animals, even if they do have several different forms of vocal expression. The most well-known horse sound is neighing; produced by the vocal chords, this sound can be heard by other horses up to about half a mile (1 km) away! When it neighs, a horse's neckline is taut and its ears point forward, with its mouth and nostrils wide open. On the other hand, horses keep their lips tight when they chirp — a common sound they make when meeting each other — and they will also use their sense of smell to get more information about the other. For a long time, it was thought that mares produced this relatively high-pitched sound more often than males, but studies have shown equal use. These observations even demonstrated that the intensity of chirping served the purpose of asserting the dominance of one individual over another. Sometimes, this sound is accompanied by the raising of a foreleg toward the sky and forward, and ears are laid flat against the neck; a horse acting this way generally doesn't appear to be very happy with the other individual it's met. Here, body language is added to auditory communication, which can take many forms, the most well-known of which is "ear language."

The ears clearly show a horse's mood. For example, when a horse's ears are lying flat and pointing downward against its neck, it's an expression of annoyance, and is perfectly identified by fellow horses as a threat and/or a form of intimidation. Aside from just the ears, many parts of a horse's body can send specific signals: the way the head and neckline are set, the raising of a hind leg (or not), slow or irregular movements of the tail, the tail raised high or flat against the rump and so on. It isn't always easy to observe these postures in a group of horses since, most of the time, they are very discreet.

Indeed, equine communication utilizes a range of signals that can crescendo given the response, the situation and/or the degree of intimacy between the individuals. If a horse approaches another, the latter may not want to be bothered, but it will not immediately bite the interloper. Rather, it will use all the sounds and body language at its disposal to warn the offender. If the intruder understands at the first signal, everything stops there. If not, the bothered horse will continue until the other moves away. We first see the flat ears, then the contraction of the nostrils and of the mouth, then the croup turning, the swishing of the tail, a few steps backward, the raising of a hind leg and, finally, the kick of a hind leg if the other horse truly insists too much. However, the hit will be a controlled one. Horses don't seek to hurt one another, and outside of fights between stallions, they're rather pacifistic animals that have hierarchies in place to avoid conflict.

Opposite: In the harem, horses may forge stronger links with certain companions — bonds that we can almost compare to friendship. These Mangalarga Marchadors appreciate grazing beside each other. They are very likely privileged partners for mutual grooming. Between them, there is such intimacy that, even if they were separated for a long time, they would probably recognize each other by smell when meeting again.

When a horse is submitted to stress, as in this kind of situation where several groups meet, it can show signs of aggression towards other horses. It will express its threat with its entire body and, as seen in this Icelandic horse, the neckline will be taut, the ears flattened against the neck and the facial muscles contracted.

HIERARCHY

Who is dominating? Who is dominated? In a harem, the stallion is the boss. The oldest mare holds ascendancy over the others and, of course, foals always defer to their seniors. In an established group, the hierarchy rarely changes once it's established, with only a few rare exceptions. These "rules" accepted by all allow the resolution of tensions that could otherwise manifest themselves into conflict. Accordingly, older members dominate younger ones, and as the latter age, they gain their place by showing respect and making themselves respected. It's important to note that horses, in contrast with other species, do not measure dominance by being bigger or heavier: in a domesticated group, a 3-foot (1 m) tall Shetland may dominate a much larger horse!

We can observe a linear hierarchy among a harem's mares. The oldest are the most respected, and the newest arrivals or last-borns must submit to them. If mare A dominates mare B, who dominates mare C, then A necessarily will dominate C. In contrast, this relationship fluctuates more within a group of single males; male A can dominate male B, but C may dominate A. There is less stability in these groups; members come and go, and their organization is rather nonlinear. One can even observe alliances being made and undone. Two stallions may unite for a given time in order to "steal" mares from an older and stronger stallion, but one of them will always dominate the other.

Ethological studies have also shown that a horse's rank within the harem does not prevent it from becoming a "leader" during certain situations: for example, if a member of the herd walks toward a water source, others may follow it. In fact, it is not rare to see harems meeting at the same water source. Several families will coexist and know each other as a result of vital needs. There is also a pre-existing order to determine which group drinks first. One herd may wait for another to finish drinking while another more dominant group may chase away the drinking harem entirely.

Opposite: Hierarchical relations within groups allow tensions between individuals to be appeased. When two horses meet, they sniff each other and establish a dominant/dominated relationship. But the dominant horse will not always be the strongest or the tallest, in contrast with many other animal species. In this case, a mini Shetland may very well exert dominance over a Friesian.

When horses meet for the first time, we can observe a number of strong reactions which can provoke certain forms of aggression. These two Icelandics shove each other without hesitation. They are not playing, and their blows can initially be quite severe. Yet this is not a fight, and it is rare to see bites and kicks in these instances.

Who dominates?

It is easy to tell who dominates who when observing this group of Haflingers. The first has an aggressive attitude with its teeth menacingly bared. The other shows a posture of submissiveness by galloping away. It does not show signs of fear, but one can see a glimmer of worry in its eyes; its flight aims to show that it accepts the dominant horse's status.

84 THE NUANCES OF FRIENDSHIP

Waltzing for dominance, these two Haflingers are gauging each other. Their necklines are rounded, their muscles are flexed, their tails swish the air and a few kicks are thrown, but no hits are landed. Horses are, above all, peace-loving animals — they particularly seek social comfort. Their intimidating postures are ideal for demonstrating strength and building consensus. Here, no conflict should be started.

Above: Horses use all of their senses to communicate. Depending on the situation, they may use auditory, visual, olfactory and/or tactile methods of communication. By sniffing each other, they use olfactory signals to recognize or get to know each other.

By touching one another, they maintain bonds of friendship. Two "friends," like these Barbs, will interact affectionately by gently rubbing each other, or by resting their head on the partner's neckline when resting.

Opposite: Horses love rubbing their heads together. The head is a delicate part of a horse's body. There is little flesh for cushioning compared to the croup or flanks, and horses show great tenderness when they scratch each other. This chestnut Barb looks like it really likes its bay partner's affections, as the yawnlike grimace shows.

> **"** The only hierarchy that counts is of the heart. **"**
>
> George Allain

As with many a learning process, these foals assimilate codes of communication by imitating adults. These two Welsh foals, a breed of pony hailing from England, test out their mutual reactions by moving their ears in different ways, by nibbling at each other with their lips rather than their teeth, and other forms of communication. The complicity forged during their childhood may endure if both ponies remain within the same harem.

Horses primarily use body language to communicate among themselves, but vocalizations are also part of their common vocabulary. This Paso Fino (opposite) appears to be combining several vocalizations, between neighing and sexual invitation, as its nostrils are dilated and its mouth is slightly open with teeth barely showing. This Mangalarga Marchador (above) is neighing. With its mouth wide open, it emits a powerful sound that can be heard up to a half a mile (1 km) away by other horses.

An Arabian Pinto neighs while exhaling, thanks to its larynx. When the air expelled from the lungs passes through the pharynx and nasal cavities, it can be modulated to be more or less powerful. The sound is high-pitched at first, moving deeper during the three or so seconds it can last.

The neigh produced by this Tinker is also called a "contact call" because it allows the horse to inform others of its presence and, according to their response, to venture over to them. Seeing the whites of an animal's eyes can indicate stress in searching for its companions. But beware, as with certain breeds, like the Appaloosa, the whites of the eyes are always visible, so it should not be interpreted solely as a sign of tension.

In the foreground, this Fjord horse is asking its partner to play by nudging with its nose. With its partner not responding, the horse becomes more insistent by pushing against its companion's neckline, with its ears slightly downward and to the side. We can also see the upper lip advancing slightly: this position of the mouth indicates that the horse is friendly and wants to play. Suddenly, its companion answers, ears turned toward its partner, and it utters a high-pitched neigh, with its right foreleg thrown forward.

It's time to play!

After a little while, these two friends will calm down. Communication between horses is a progressive business: their posture, movements and so on will always crescendo and wane. Horses often finish their games with mutual grooming, as if to seal the bonds of their friendship. This contributes significantly to horses' physical and mental well-being.

PARTNERSHIPS

—

The first partnership is the group itself. Horses' gregarious instincts encourage them to maintain social relationships with their companions, above all for the purposes of survival. Horses must always be on alert for predators, and so they must always be watching their surroundings. It takes more than one horse to do the job right. In a group, they can alternate roles, watchers become sleepers and then the reverse. It also becomes easier to manage food and water supply by using the environmental knowledge of the "elders." For example, in times of drought, an older mare may remember a different watering hole from the usual one. It is also easier to protect foals in groups, and so ensure the reproduction of their species — a truly genetic instinct.

The second partnership, and the most easily observable, is mutual grooming. Just like monkeys delousing each other, this interaction is both necessary for hygiene and well-being that provides an opportunity to forge bonds between members of the harem. Horses only practice grooming with companions with whom they have an affinity. Similarly, horses generally have a preferred partner with which to rest; a certain form of intimacy seems necessary to allow for more serenity. To groom, partners stand head-to-tail with each other and nibble each other's neckline, withers and all along the back, right down to the dock. In springtime, when the winter coat is molting, this activity is particularly useful to get rid of dead hair, especially in places horses can't reach by themselves without rolling on the ground. However, it appears that grooming also serves the purpose of reinforcing social bonds — partnerships that provide tranquility and relaxation. Scientific studies have also shown that the heart rate of domestic horses slows down when one pets their withers. Similarly, grooms have noticed that certain horses greatly appreciate when their dock (the tail's base) is gently massaged.

Opposite: The adorable faces of these two Icelandic horses will charm any observer. In fact, both of them are also observing their surroundings with interest. Their ears point forward in order to hear better, and their eyes are captivated by the object of their attention. They stand side by side as if to assert their mutual confidence and to show their adversary the potential of their collective strength.

These two American Quarter Horses have no problem sharing the same patch of grass. When two horses are fond of one another, they become true partners in life. As such, they like grazing together or, like the Mangalarga Marchador (at right), to share in tender affection before or after resting.

It has long been thought that the oldest mare leads the herd from one pasture or water source to the next. Recent ethological studies tend to demonstrate that any horse can become the leader for a given activity. The lead horse in this group of galloping Connemaras could be one of the lower-ranking members of the hierarchy — or not.

Mares and foals are naturally privileged partners. These Welsh foals have nothing to fear — their mothers are watching out for them. Even when they are calmly grazing, mares will react and raise their heads at the slightest noise. They scout the horizon to make sure that their offspring aren't at risk. Their foals imitate them, thus learning the codes of communication within the harem.

Resting time

These Mangalarga Marchadors' rest has been disturbed while they were lightly sleeping in "cow position" (with their legs folded under them). They now appear to be particularly wary, in contrast with the group opposite, resting in an upright position. Indeed, horses can sleep standing up thanks to a knee-locking mechanism. They often alternate between flexing one hind leg and another, and they keep their eyes half-closed, with their bottom lip slightly hanging, and their ears placed backward and to the side. Resting upright gives a significant advantage to the horse if it needs to quickly flee from predators. It is absolutely indispensable to the survival of this prey animal.

This Icelandic foal seems to have been lulled into the deepest of sleeps. It is lying with its flank to the ground, its head and limbs sprawled out over the grass. Deep sleep only lasts a few minutes for horses, since their survival depends on quick reaction times. An adult horse only sleeps deeply for about two or three hours per 24-hour period, while foals require more resting time as they grow.

Muscles are completely relaxed during deep sleep, which allows for physical recuperation to take place. The mouth is generally a little ajar, just as with this Fjord foal. Horses often produce moan-like sounds while their bodies twitch slightly, especially around their closed eyes. In this case, horses are in the REM sleep phase and are dreaming.

Members of the harem watch out for danger while their companions sleep. The German horse opposite and Pinto Arabian above are both on guard. Their ears are pointed up, their expressions are particularly attentive, and their necklines are raised — they are very wary of their surroundings in order to be able to alert the group in case of a threat. They use their sight and hearing during the watch as well as their sense of smell, all to better detect a potential predator.

Next pages: We can see in this trio that muscular tension is not the same for each and every member of the harem. The horse at rear seems to be quite vigilant indeed, even though it is lying down on its stomach. The other two, however, are more relaxed. Their necklines are lower and their ears droop slightly to the sides — they are slowly falling into light sleep. They may even rest their chins against the ground when lying this way to get a better rest.

These two Mangalarga Marchadors are not biting each other's withers, but rather are contributing to the well-being of their partner through what we call mutual grooming. This equine behavior has three functions: maintaining good skin hygiene, reinforcing bonds between partners, and bringing tranquility. In terms of "skincare," grooming allows the horses to get rid of dead skin and hair, especially during the spring molting phase (when a horse sheds its winter coat), and it also helps to keep certain parasites at bay.

A solitary horse can clean its coat by rubbing against a tree, rolling on the ground or even nibbling itself, but it will not be able to reach certain parts of its body alone. Moreover, mutual grooming serves a particularly important social function. Observations have shown that this behavior is seen between two horses that share a true affinity for each other, and studies demonstrate that scratching each other's withers (the favorite spot) even contributes to a reduced heart rate, and as a result reduces muscular and social tension.

Mutual grooming

3

GALLOPING TO
FREEDOM

Although horses possess the same senses as humans, they do not share the same sensitivity, nor the same perception of their environment. Their behavior may be surprising, since we don't perceive the same information as they do. Between their experience and ours, there is a world of difference! And this should be understood in both the literal and figurative senses, since horses have much better vision than humans do. When stopped, perfectly still and looking straight, a horse does have two blind spots directly in front and behind it. However, with eyes on each side of its head and a long flexible neck, a horse needs only turn its head for its vision to become nearly panoramic. A horse may sometimes be taken by surprise from the back and give a reflexive kick to protect itself, but this remains a rare occurrence, since it's always aware of its environment thanks to its highly developed senses.

As a result, its eyes are constantly watching for the slightest movements in its surroundings. This is useful for spotting far off predators that may want to slip into the high grasses to attack. When a suspicious object goes through its field of vision, the horse may step or flee. Like other prey animals, horses prefer to protect themselves first, and then find out what the danger is once they are safe. They may also be startled when entering a wooded area. A horse's eyes have more trouble adjusting to changes in luminosity, and although their night vision is better than ours, they cannot pass from light to dark as easily. Similarly, a horse can be unsettled by alternating light and dark spots when on a path between trees, which may make it stand still or want to turn back.

To better understand equine behavior, humans must not only observe a horse's entire body and how it moves, but also its environment. Whether on foot or on horseback, one must always be ready for the animal's potential actions. It is important to give a horse enough freedom of movement so that it can readily discover an object, as it will use all its senses to do so. When faced with the unknown, it will first observe closely (horses see better from up close), sniff ("Does it smell good? Is it edible or not?") then, if it is reassured, it will lightly touch the object with its lips or its foreleg ("What kind of noise does it make if I touch it"), and even lick and taste it ("It looks like that bag of hard bread they bring me sometimes. Is there any in there?"). Once its curiosity is sated and its fear appeased, the horse will continue along.

Horses particularly want to do this since they love wandering. Moving around, generally at walking pace, is invaluable behavior in the wild, and horses will only trot and gallop when fleeing or playing. This allows them to go from one pasture to another or to head for water sources, their movements being essentially tied to nourishment. Compared to other species (mostly predators), horses do not have territory, but require vital areas where they can find food, water and shelter from the weather. This area varies in size according to the region they live in; in an arid climate, the scarcity of food necessarily implies a larger zone.

Opposite: This German Saddle Pony's forelock is speckled with snowflakes, and a few more hang on his vibrissae (long tactile hairs around the nostrils and mouth that allow a horse to better sense the objects it encounters). Thanks to fur that thickens when sunlight wanes and temperatures fall, this horse doesn't suffer from the cold of winter.

This group of Curly horses, with two adults flanked by two youths, can be seen galloping across their lands, their senses fully alert. The brown bay horse's gaze, seen on the left of the photograph, as well as its forward-facing ears, show that it is very aware of its environment. The two younger horses behind follow along, their attention more seemingly focused on the attitude they must show their elders.

FINELY HONED SENSES

Humans are soundly beaten by horses in terms of hearing. Horses can not only pivot their ears in every direction in order to identify the source of a noise, but they can even hear at ultrasound frequencies. So, if car or motorcycle drivers equip their vehicles with ultrasound warning signals to move animals out of their way, this poses certain risks for horses who may be calmly grazing in a pasture nearby, and would suddenly be startled for apparently no reason. If the ultrasound signals stop, an attentive observer will see the result, with the horse quickly calming down. But like any good scientist, an observer must test out different situations in order to verify the initial hypothesis. Their observations must be informed by knowledge acquired through experience. In the previous example, few domesticated horses are scared of vehicles, but the sudden appearance of a motorcycle or the rumbling of an engine can scare them, just like the smell of a wild boar in the woods. A horse's sense of smell is quite honed.

In fact, they use their sense of smell to recognize each other. By licking her newborn just after giving birth, the mare permeates herself with the smell of her foal, and she will not (or will rarely) accept another foal feeding from her. Whether it be licking or touching, tactile contact is essential for horses. Combined with sight and taste, it is an essential tool for communication with others, for exploring its environment and especially for feeding itself. It is not rare to see a horse refuse to drink dirty water, for instance; it knows that between the bad smell, the presence of algae and the awful taste, the water will likely make it sick rather than hydrate it, and it will abstain until it finds a source of drinkable water. That's right — horses have taste! Just like humans, they can distinguish sweetness, bitterness, acidity and saltiness, and each horse will have its favorite tastes. Some will enjoy the carrot greens as much as the root, but they rarely go crazy for salad; horses may be herbivores but they like more than just greens.

Opposite: Do we even have to note that this is a Curly horse? Its frizzy mane and coat are a dead giveaway. The hairs inside its ears form a billowing down that does not impede its hearing, which is as acute as all other horses. However, this bushy ear hair is definitely an asset when it comes to hygiene, as it prevents dust and other small particles from getting in.

Horses' ears are very mobile. They can turn them almost 180 degrees in every direction. Both ears operate independently from the other. When it hears a noise, a horse turns its ears toward it, and if the sound is a familiar one, it stays put and goes back to what it was doing. If it is not able to identify the sound, it will turn its head to visually check whether there is danger or not. When it starts moving, like the galloping Pinto opposite, its ears will generally face front, except when fleeing and listening to a sound behind it. The rather conical shape of the ears, like a kind of little conch shell, increases the efficiency of the auditory system by enhancing resonance.

The hearing, sight and sense of smell of these three Morgans is on full alert and, as equine communication dictates, they are perfectly synchronized in terms of their attention. Their ears stand almost straight up and face forward in order to capture the slightest noise, and their eyes are fixed on the source of the noise.

Thanks to their sensitive sense of smell, horses can detect the scent of a predator before it gets too close, especially if they are upwind. There is a hitch though: if there is too much wind, their hearing can be a little "scrambled," and they will have to place more faith in their other senses.

GALLOPING TO FREEDOM

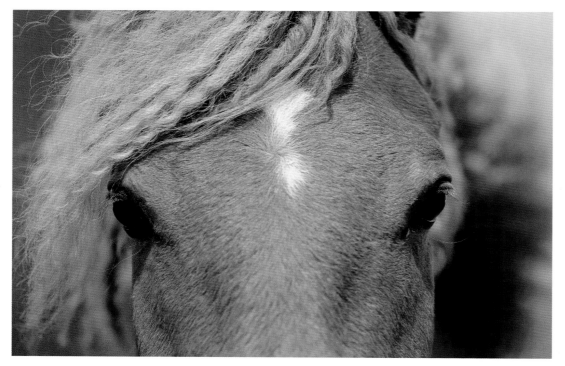

Left: Horses' eyes are located on the sides of their head. This is the case for many herbivores and prey animals whereas predators (wolves, lions, cougars and so on) have their eyes at the front of the face. This ocular position allows the horse to have a near-panoramic view of its surroundings. Horses have both monocular and binocular sight. When its body is perfectly straight and immobile, there are two little "blind" spots in front and behind the horse. It only needs to turn its head slightly, thanks to the flexibility of its neck, in order to widen its field of view.

Opposite: A horse can also adjust the opening of its pupil and eyelids according to the brightness around it. However, it takes a little time for it to do so. If this gray Connemara goes into the shade too quickly, its eyes will take a certain amount of time to become accustomed to the lesser light, and the horse will be somewhat blinded for a few moments, and may even stand still for a moment while its eyes adjust, and to make sure it is not going anywhere dangerous. This sensitivity to light also has advantages: horses have better night vision than humans, although less so than nocturnal animals like cats.

FINELY HONED SENSES

The flehmen response

A horse's sense of smell doesn't allow it to detect pheromones, the chemical substances produced by glands, but its vomeronasal organs do. These organs are situated under the palate and on each side of the nostrils. In the flehmen response, a horse curls back its upper lip, and this blocks the nostrils so that the air breathed in goes directly to the vomeronasal organ. The facial movements are characteristic: the horse stretches its neck and head upward, breathes in, then curls back its upper lip, baring its gums. The eyes are pulled backwards by the jaw's contraction, and we can see the whites of the horse's eyes.

> "In each sense are the five others."
>
> Juan Ramon Jiménez, Spanish poet (1881–1958)

Previous pages: The horse's sense of smell, aside from the flehmen response, has not been studied very much, but observations tend to prove that it's more developed than that of humans. Thanks to their olfactory senses, horses recognize each other within a group, and can evaluate food before tasting it, for example. This Icelandic horse on a lakeshore may be studying the smells left by horses there before it, or it may be verifying whether the ground, though damp, is suitable for a nice roll.

This page: Contrary to long-standing assumptions, horses do not have thick and insensitive skin. The proof? It can feel a fly landing on any part of its body and twitch its muscles in that specific area in order to repel the intruder. Its head and limbs are particularly sensitive, since the flesh there is less thick than on meaty parts like the croup.

Nonetheless, and as part of good hygiene, these parts must be scratched and rubbed in order to get rid of dead skin and parasites, and, in this case, horses may use "tools" like a tree trunk. This stable prop allows the horse to put its full weight against it, and its coarseness is perfect for getting rid of dead hair.

When there's nothing to scratch against, and when there are no friends around who want to groom, horses may scratch themselves by using their hoofs, just like this Icelandic pony (left), or its teeth, like this Arabian purebred (opposite). This technique, however, is less efficient: the foal must take a position on three legs, and, once it reaches adulthood, it will only be able to repeat this maneuver with lots of balance and flexibility. Scratching itself with its own teeth is also much less gratifying than a mutual grooming session with another member of the harem.

FINELY HONED SENSES

This page: Touch allows the horse to understand its environment. The Fjord horse (left) and Mangalarga Marchador (below) are curious about this tree, its spring flowers and summer foliage.

By approaching the branches this way, horses are careful not to hurt themselves — especially their eyes. They are helped by their vibrissae: long tactile white hairs around the mouth and nostrils and also around the eyes. When they are in contact with an object, these hairs inform the horse of its proximity to an object so that the horse doesn't bump into it.

Opposite: Horses are curious by nature and, after having sniffed and touched this piece of wood with the tip of its nose, this Mangalarga Marchador is not afraid to pick it up with its teeth. In contrast with other animal species, like monkeys for example, horses will not use a piece of wood as a tool, and they will not confuse it with something edible. This horse will most likely use it to play with. Light and small, this tree branch can be thrown, chewed, cracked underfoot and so on.

These Haflingers are enjoying some good hydration at a watering hole. Before dipping their mouths to drink, horses take several precautions. First, they test the surrounding ground. They must not be trapped by the land they're standing on; and venturing out into marshlands, where they risk getting stuck for good, is out of the question. They verify that the water is drinkable by smelling it. A foul odor will immediately deter this duo from drinking. Finally, their downward-hanging necks not only allow them to approach the water and drink, but also to use their eyes to verify that no danger is present. Horses will always lower their heads to inspect something more closely.

Depending on their home region and the weather conditions there, horses will judiciously adapt their feeding habits. This Mangalarga Marchador has surely learned that scratching at the snow with its foreleg or snout might sometimes yield some grass to eat. Of course, grass is not very nutritional at this time of year, but it will still give the horse some sustenance. Horses have ingenuity and adaptability when it comes to feeding themselves. Certain rustic pony breeds have developed thick hair around their noses in order to protect themselves when nibbling at the leaves of spiny bushes.

These two Fjord horses certainly don't need to worry about feeding: the grass has a magnificent green color and seems quite tasty. If they were in Camargue, would they develop the ability of eating algae from beneath the water as do the small gray horses of that Rhône region? In any case, if these two are mares with foal or are gestating, they will spend that much more time grazing, since they require the energy to develop the fetus and produce milk.

Horses' feeding habits are nearly the same everywhere and, whether it is this Icelandic horse (opposite) from northern Europe or this Haflinger of central Europe, grass is their primary food. As non-ruminant animals with small stomachs, horses must graze for a long time in order to live. On average, horses spend about 15 hours a day feeding. But they don't stay still the entire time: they slowly advance, gradually progressing toward the best tufts of grass. In prairies, horses eat different varieties of grasses, which helps them gather all the essential nutrients they require and implies that they will often move from place to place.

FINELY HONED SENSES

AT THE HORSES' PACE

—

In contrast to many other species, horses are not territorial animals. They don't fight over space, but rather share areas where they can find all their necessary resources — water, food and shelter. They share these zones with other equine family groups, seldom meeting and only in such places as watering holes. Some groups will avoid each other entirely in order to limit the possibility of any confrontation. However, scientific studies have shown that there is no single correlation, with horse behavior differing according to region. If food and water are abundant, there is obviously less of an issue with several groups coexisting, since there is no conflict over resources.

To go from one pasture or watering hole to another, horses move along paths that they forge from regular use. Along these trails, stallions have the habit of dropping their dung on top of that of their competitors. These piles give potential adversaries a wealth of information about the stallion. The harem moves at walking pace. The other two gaits, the trot and the gallop, are reserved for seduction rituals in the case of the stallion, as well as fleeing and playing.

These behaviors are shared by all types of horses. Of course, foals are the most playful, sometimes even playing with the stallion, and mares tend to be more calm in general. This habit of playing proves that horses have a primal need for movement, whether wild or domesticated. Kicks, leapfrogging, sudden sprints, rearing and other movements are practiced by all horses.

Horses play when they feel good and safe, and we can observe that they nudge, nibble and chase each other whenever they are excited, without showing threatening or aggressive behavior to their companions. Indeed, when mimicking a fight, a young male's ears are not laid flat against its neck, but stay pointed forward, a sign that there is no serious aggression toward its partner.

Opposite: This Curly stallion has a magnificent leopard coat. Its ears and eyes point toward the photographer, who it probably noticed quite quickly in its vicinity. He, like his fellow horses, do not have a territory to defend. They require only space where they can find food and water.

Even a free horse, like this piebald Tinker, is attached to a certain territory or, as in equine terms, a "vital space". The quality and quantity of food available will vary according to geographical region: a harem can live on spaces up to a few square miles in size. Horses do not feel the need to expand their territory unless they need more food or water, since they have no reason to go exploring otherwise.

Logically, when a group of horses lives in a region where food is rare, like a desert, its vital space increases and it will sometimes travel long distances to find another pasture or watering hole. Where the landscape is richer, grass more abundant and water more available, the herd will have a smaller vital area. Horses like these Arabian purebreds in a rich meadow, will travel less distance to reach each area necessary to their survival.

Each equine group adapts to the land on which it lives, and they will also adapt to the changes in weather in regions where seasons are clearly distinguishable. In summer, the sun burns the grasses and the vital area may correspondingly change or expand. To survive, horses like this Peruvian Paso also need sources of water that don't dry up.

A sheltering tree shade caresses the near immaculate coat of this Tennessee Walking Horse. A harem requires shelter, hedges and trees in its vital area. In summer, horses invariably require shade. Even more than wind and rain, horses dread the hot, intense sunlight and stinging insects like horseflies.

Next pages: These two Arabian purebreds are taking advantage of the shade by spending some quality time with one another. They are rubbing each other's heads together. Judging by the shape of their bodies, it could be a stallion to the left sharing a moment of affection with one of his mares.

Horses feel safer when they are used to a certain watering hole, and may sometimes venture farther into the water. These two Morgans seem perfectly at ease in this pond. More than just drinking, they seem to take pleasure in the bathing itself. Not all horses have the same confidence, and they will often prefer to stay on the banks of a water source, especially if the liquid is not clear enough to see the bottom and to check that no danger lurks there.

Horses often head for particular areas of their territory at specific times. In the summer, they'll look for shade and relief from insects under trees during the day, only heading toward their watering holes in the morning or in the evening. Access to a water source is essential to the harem's survival: each horse drinks an average of about 10 gallons (40 L) of water a day, and sometimes more when the temperature is very hot.

Next pages: These purebred Lusitanian mares and foals are used to high temperatures in their native land, Portugal. If the sun's rays are not too intense, they may leave the trees' shade, but they will still reduce their activities; they graze, of course, but they'll avoid romping around until the temperature has cooled down. We can also observe the foals sheltering themselves from harassing insects by putting their heads under their mothers' tails, using the long hairs as natural fly repellent.

In fall, as the leaves turn yellow and red, horses prepare physically for the coming winter. The seasonal molt will vary according to the breed. This gray purebred Arabian at left is a "surface-blooded" breed that has inherited the genes of its ancestors which hailed from arid climates. Accordingly it has a less dense coat than its companion, a Haflinger, a rustic Austrian pony that can be identified by the thickness of its coat. The Mangalarga Marchador opposite protects its neck quite well with its long mane.

Southern breeds, though used to milder climates, can adapt to much harsher conditions, even though these are quite rare in their native countries.

As proof, this chestnut Peruvian Paso (left) and this gray Paso Fino (lower and opposite) seem quite comfortable in the snow. They do not appear bothered by the snowflakes hanging off their forelocks and manes, and they appear quite at ease when galloping on the powder.

Next pages: This superb Peruvian Paso looks to be enjoying the cool, dry air of a winter day. Its bay coat — recognizable by its brown hairs and mane, and its black lower legs — contrasts wonderfully with the sky-blue glacier. If proof was needed that winter is a pleasant season for horses, this magnificent galloping horse should suffice!

> If you want to go fast, walk alone; but if you want to go far, walk together.
>
> African proverb

Horses generally walk when they are moving within their vital area. The walk is a so-called symmetrical four-beat gait, and we can see from this piebald Mangalarga Marchador that the horse first places its back right leg, then its front right, then the back left, then the front left, and so forth. It advances at about 4 mph (6 km/h).

Opposite: This Friesian is trotting, a gait generally reserved for play time. Trotting may also be used by a stallion when parading his strength before an adversary or when trying to seduce a mare. Horses travel at an average speed of about 9 mph (15 km/h) when trotting. It's a jumping and symmetrical two-beat gait: the horse plants the front right and back left legs together, then the back left and front right. The horse moves its legs jointly in diagonal pairs.

The gaits

Previous pages: This Paso Fino is galloping. This is the fastest gait, with an average speed of 12 mph (20 km/h), peaking at more than 13 mph (21 km/h). It is an asymmetrical jumping and rocking gait (the horse gallops on its front left or right leg) with three beats and a projection phase. This may seem complicated, but a keen observer can see that the galloping horse puts down its back right first, then the back left and front right (the diagonal pair) almost simultaneously, then the front left. The rear legs push all four legs off the ground (the projection phase) and the cycle starts again.

Galloping is reserved for fleeing and playing. Is the gray Spanish purebred (opposite left) fleeing from the Friesian right behind it? We can see that the latter has its ears slightly backward-facing and a hardened look, its face taut. Its companion's rather serene appearance suggests that it is not panicking one bit. The chestnut Peruvian Paso above, on the other hand, is playing alone. Galloping really is the best way to unwind.

Jumping and landing on all four feet just in time to throw its legs back for a good kick, this Noriker leaps right back into a disorderly gallop, throwing its forelegs forward like a boxer fighting an imaginary opponent. In the wild, these types of crazy antics occur when playing alone.

Even moderately deep and sandy ground does not slow down these two Barbs, a North African breed. They alternate between sudden sprints and kicks, sending plumes of dust flying. Horses show their full enthusiasm, their raw power and their absolute beauty when performing these crazy antics.